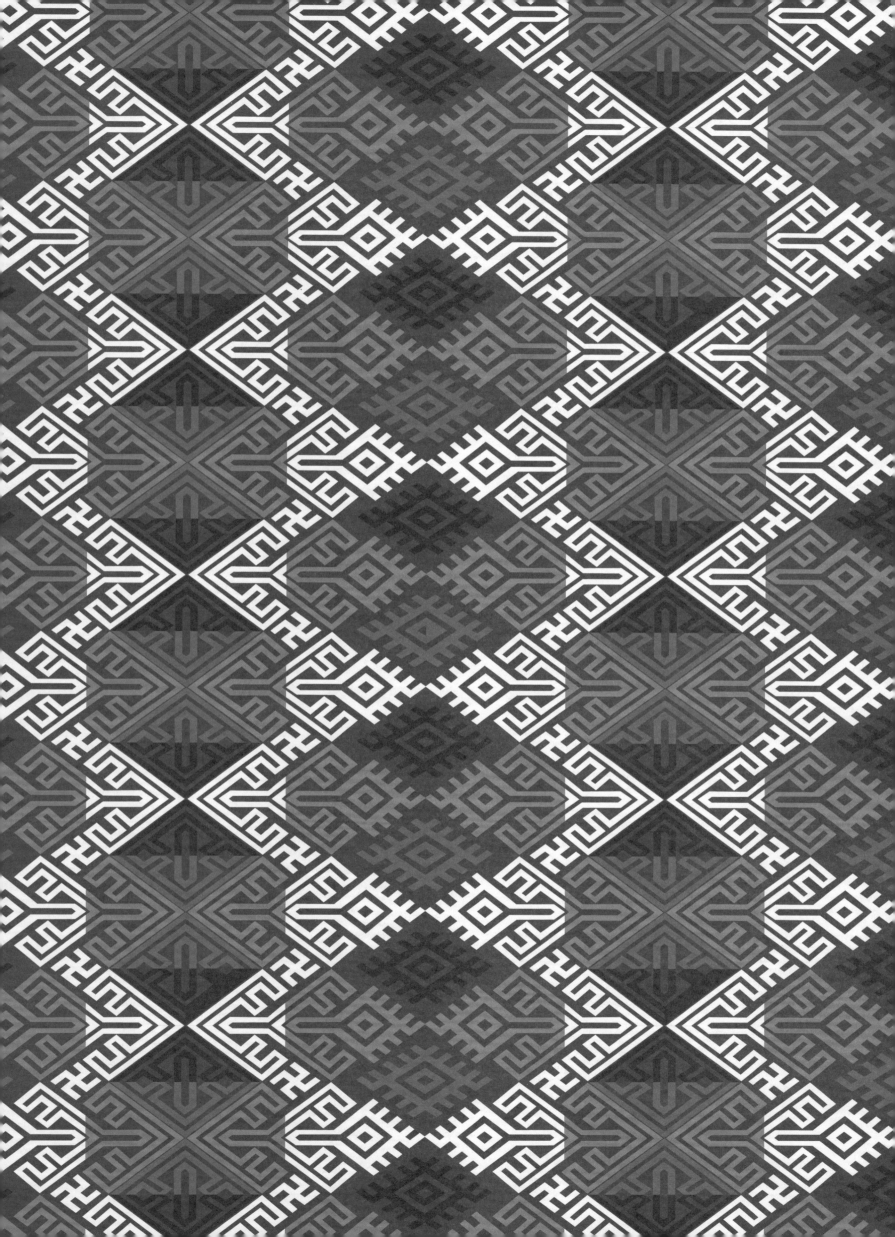

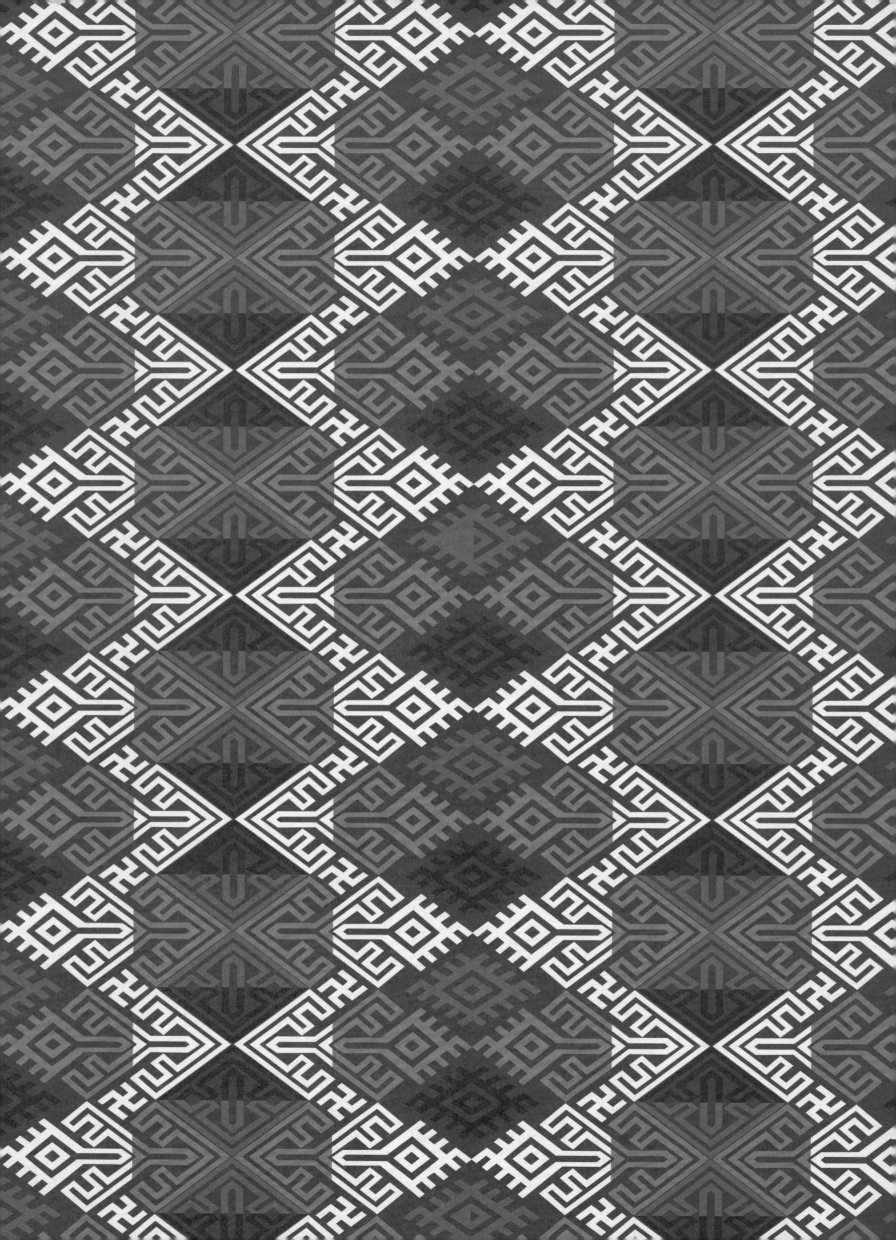

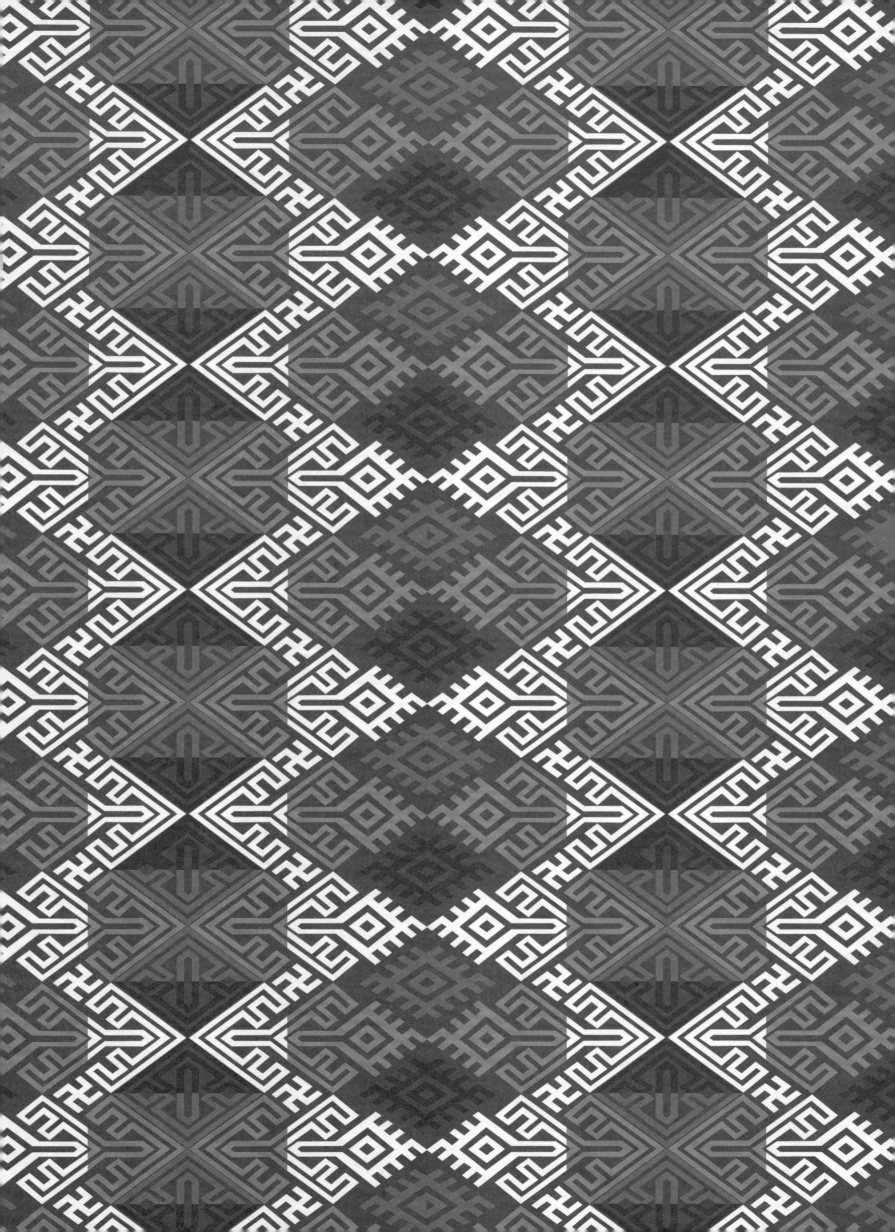

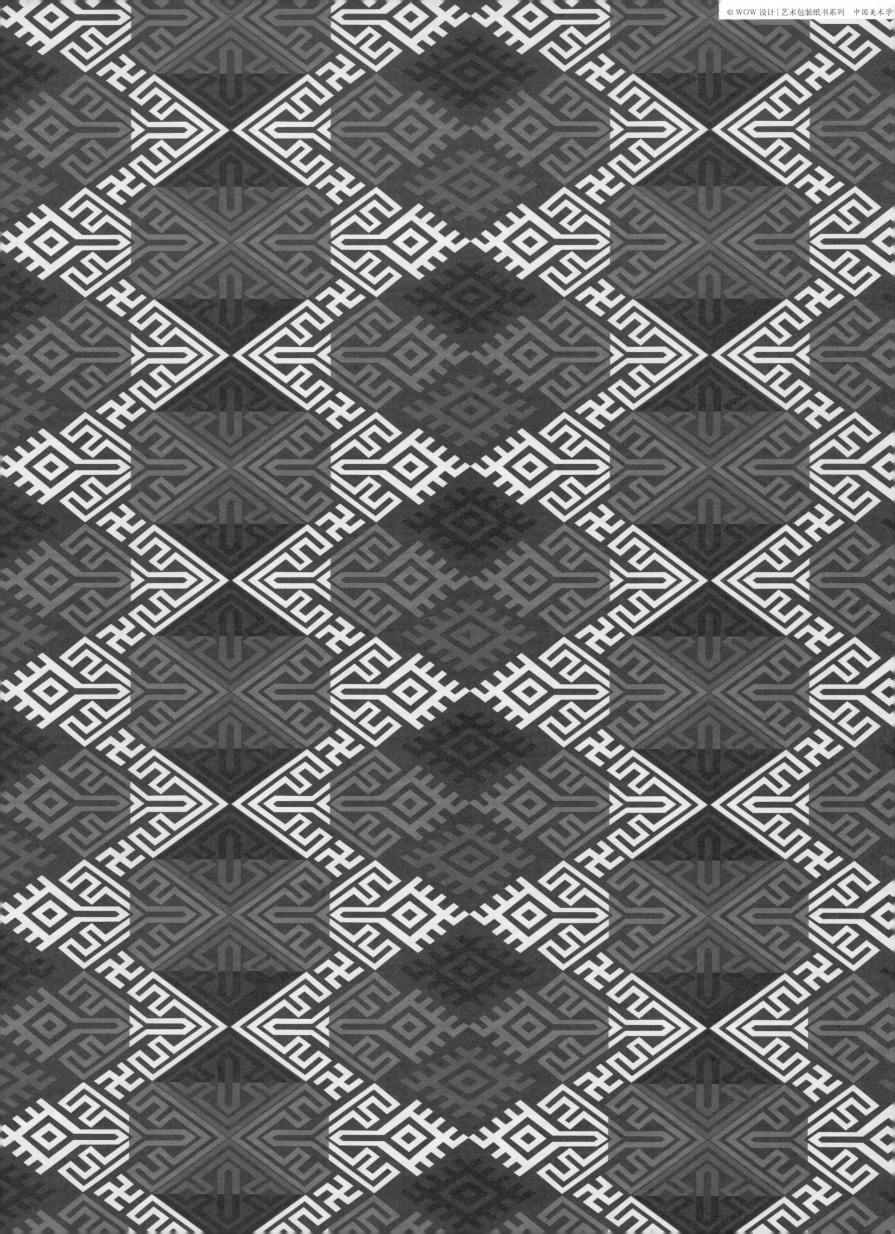

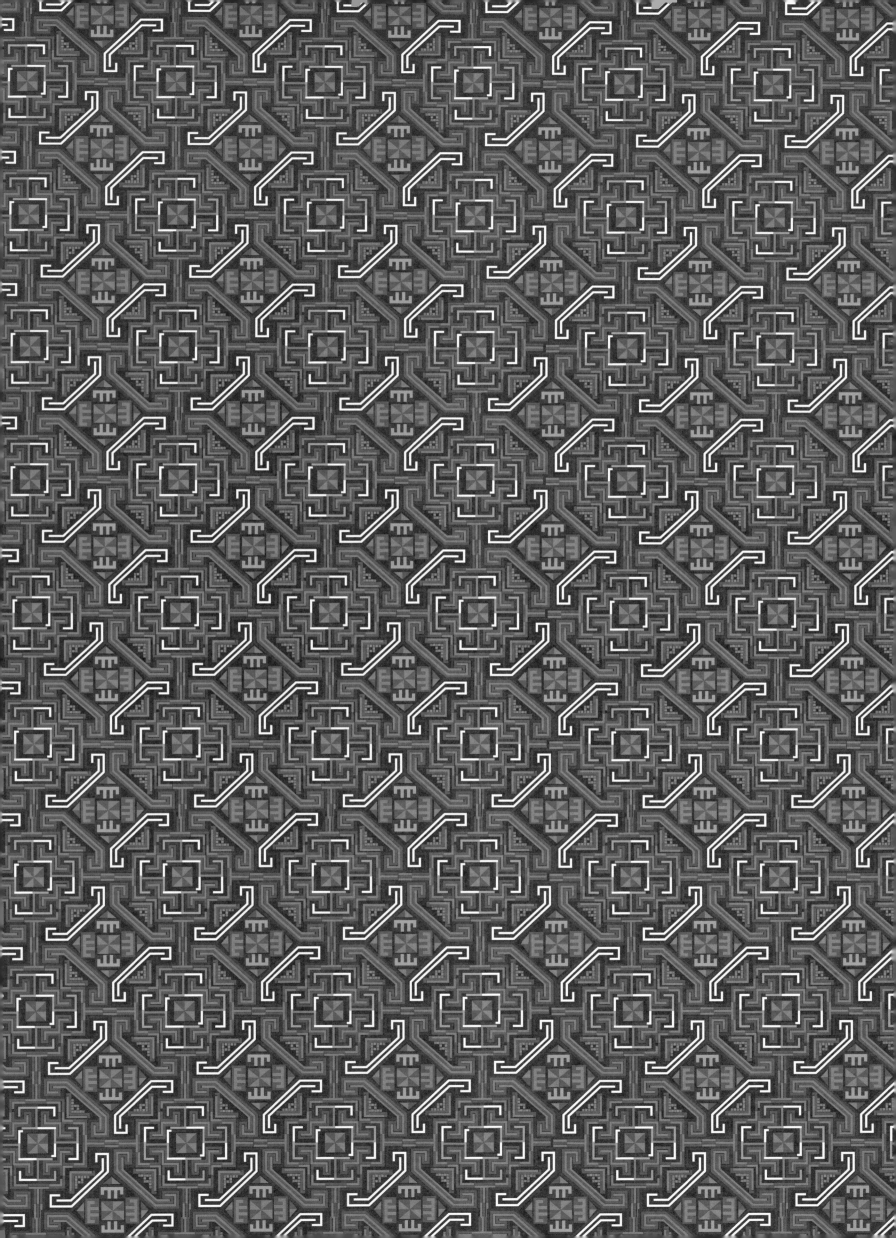

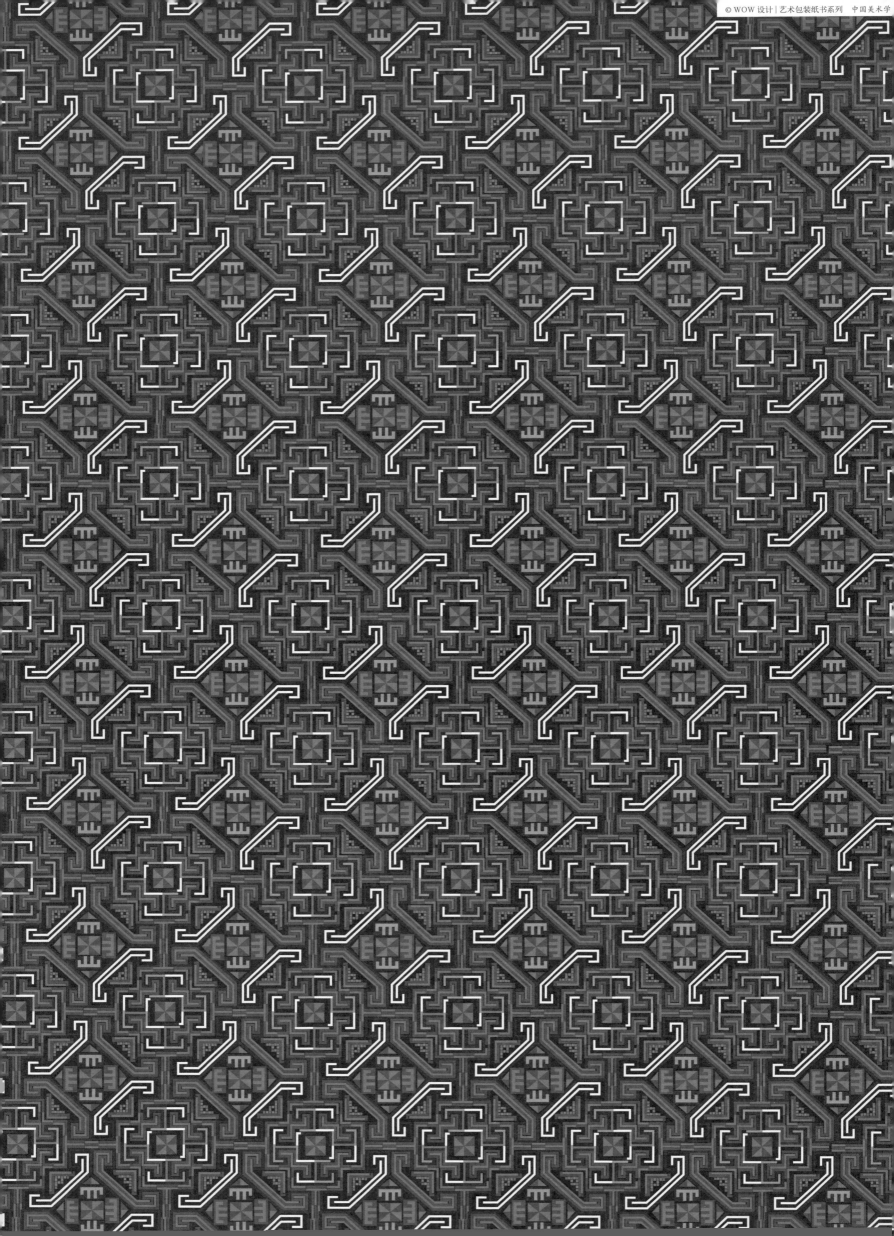

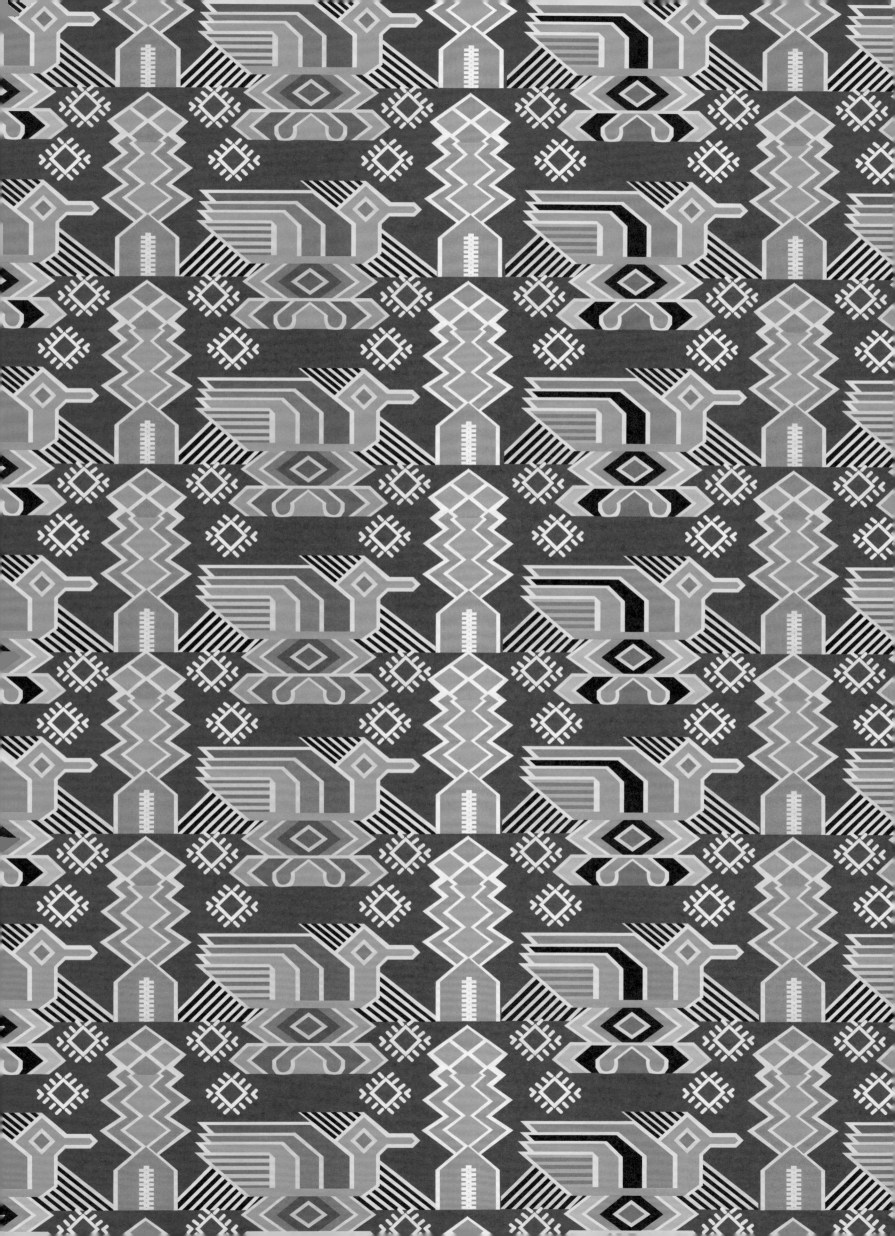

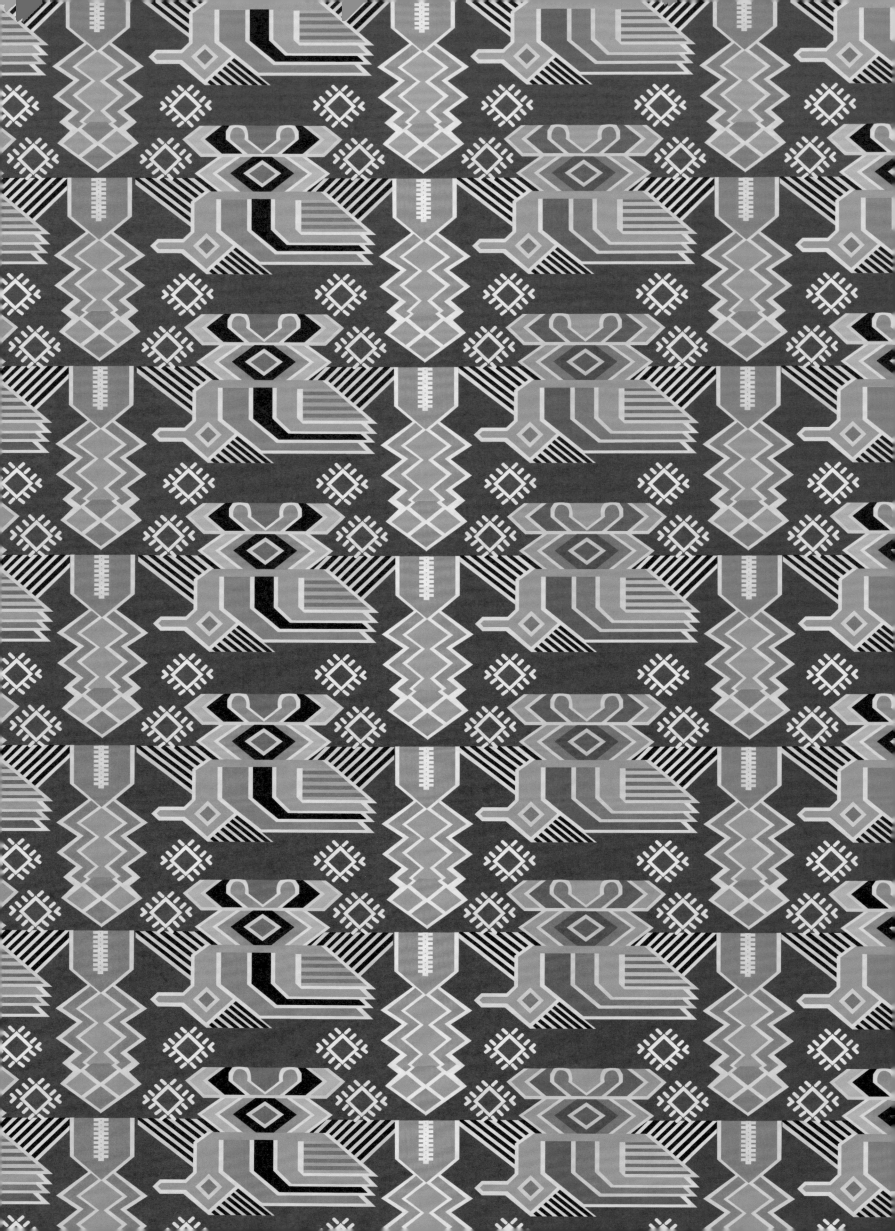

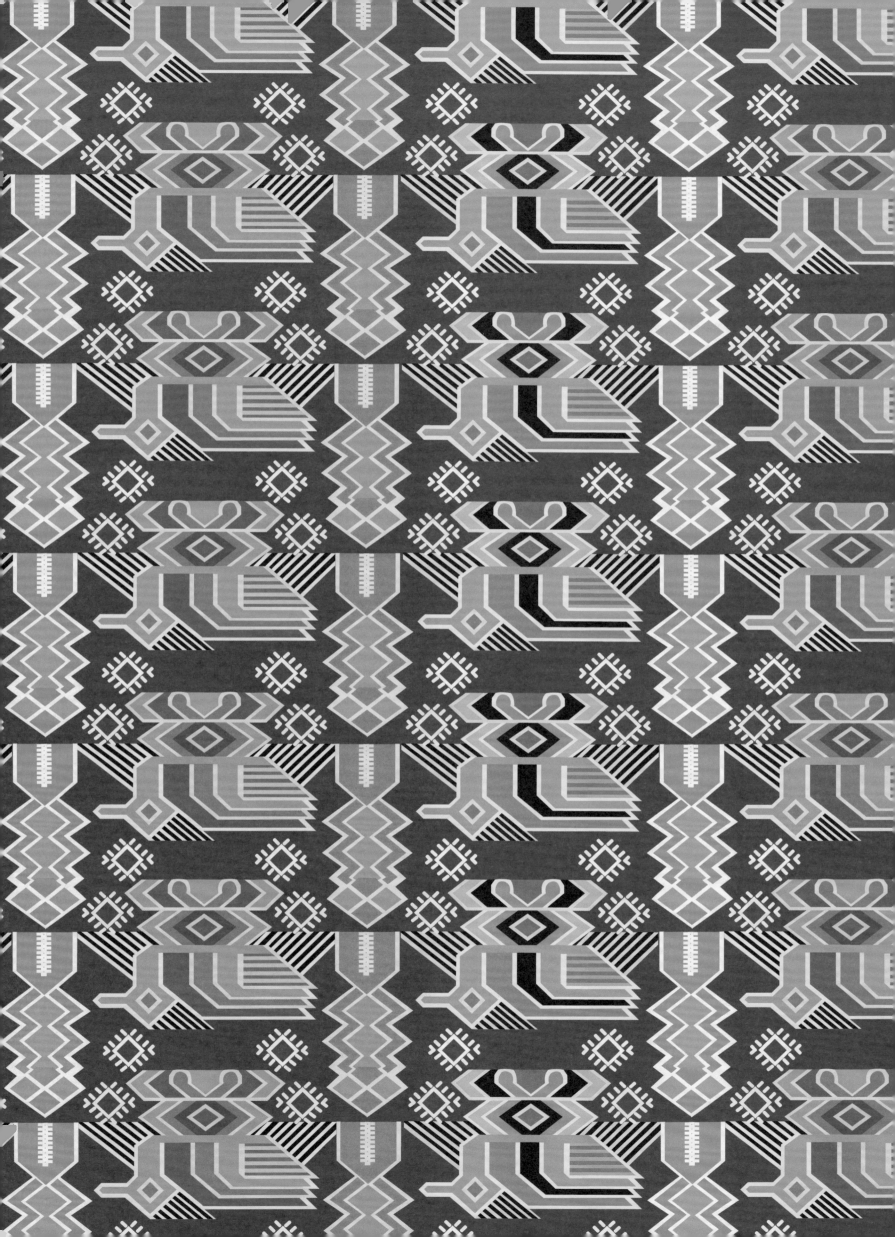

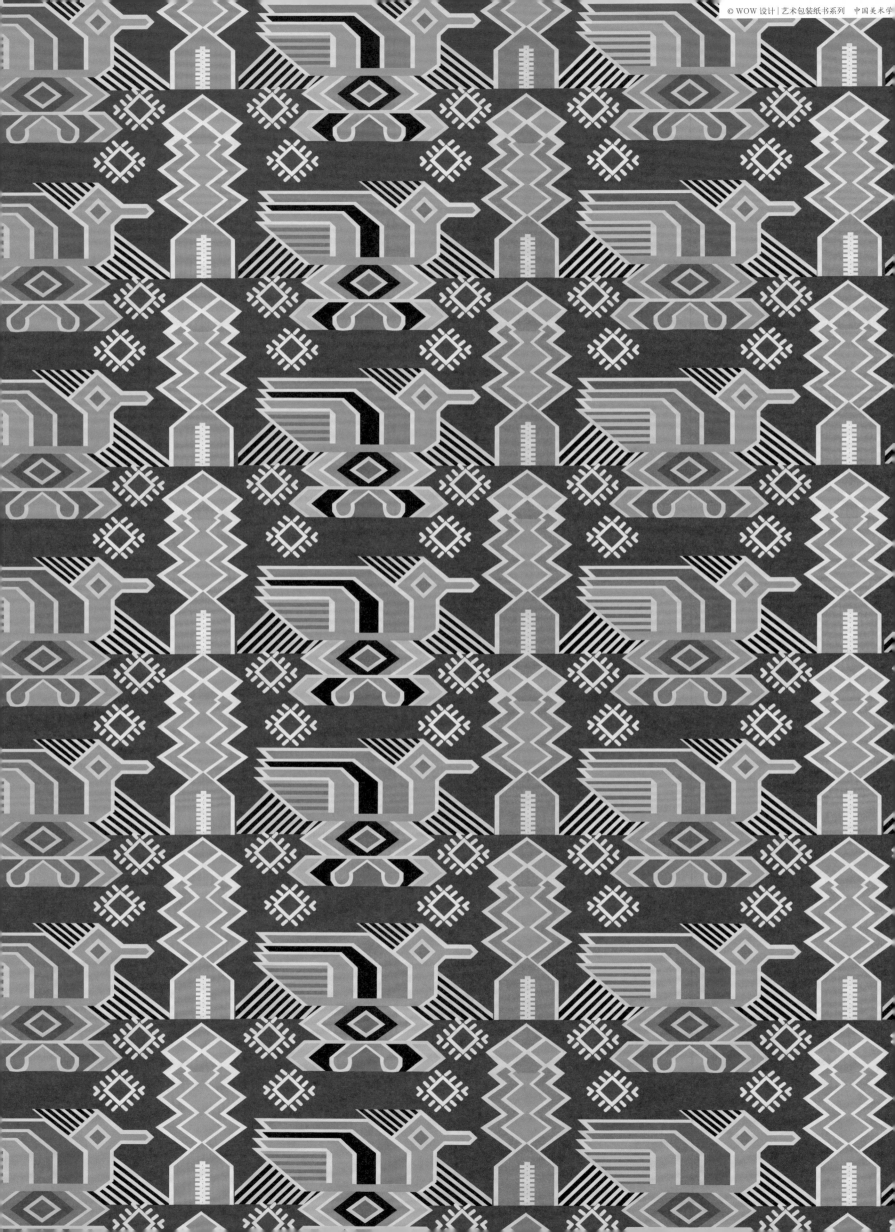

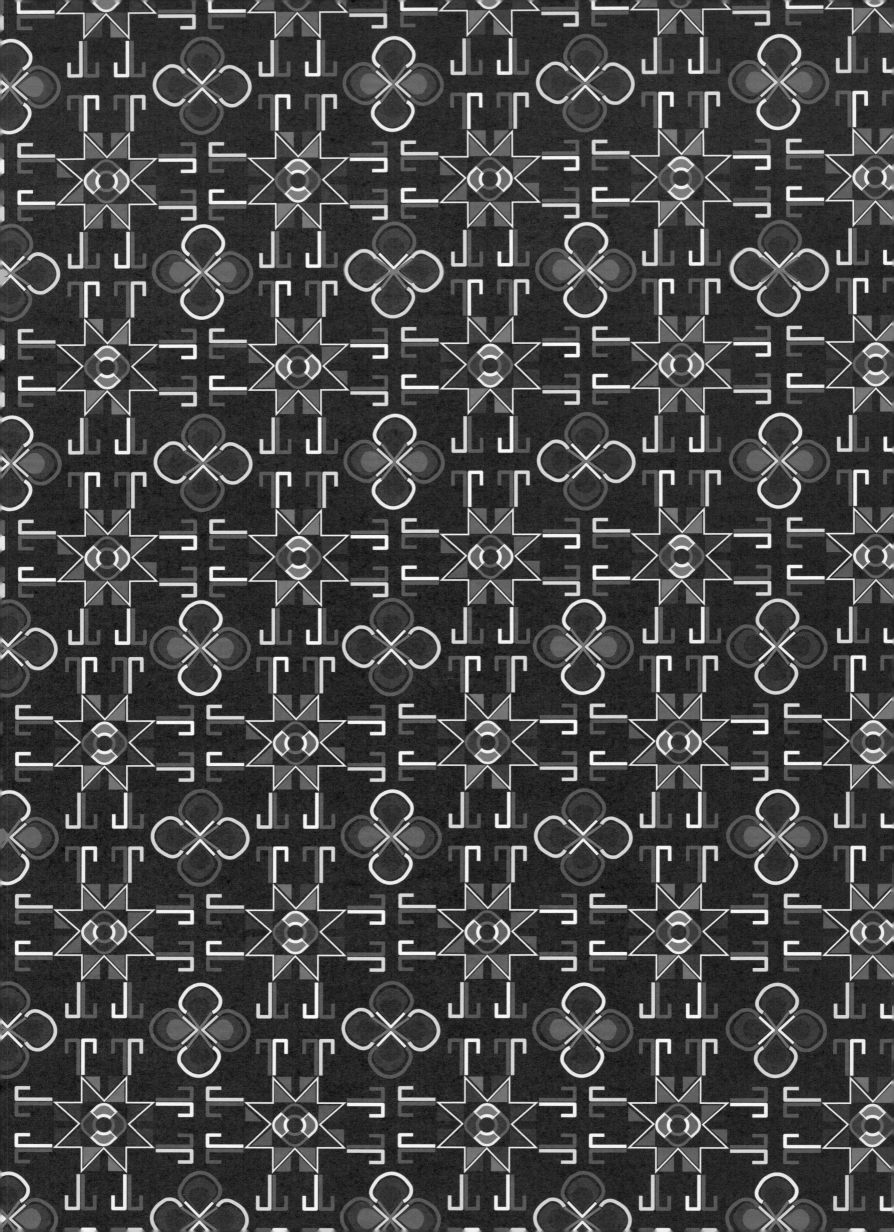

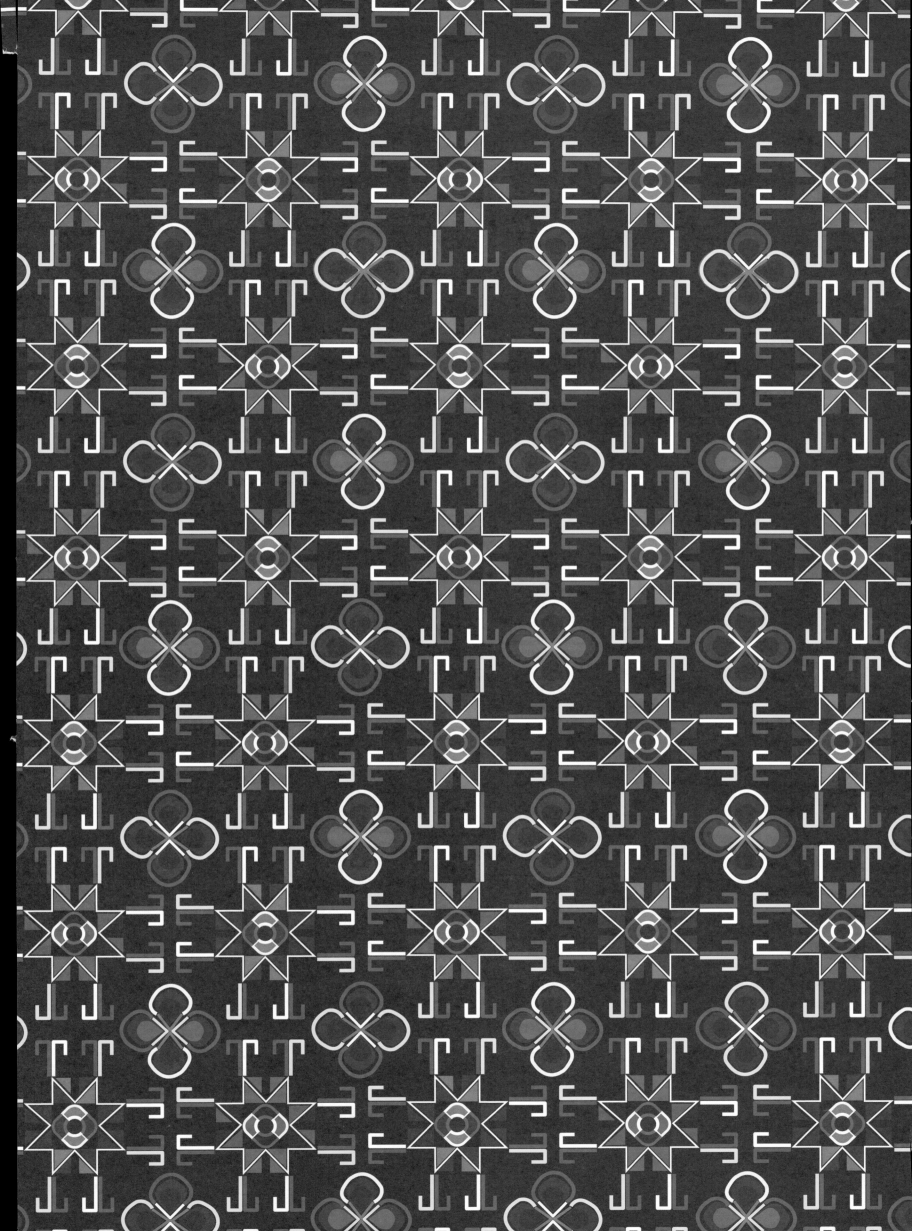

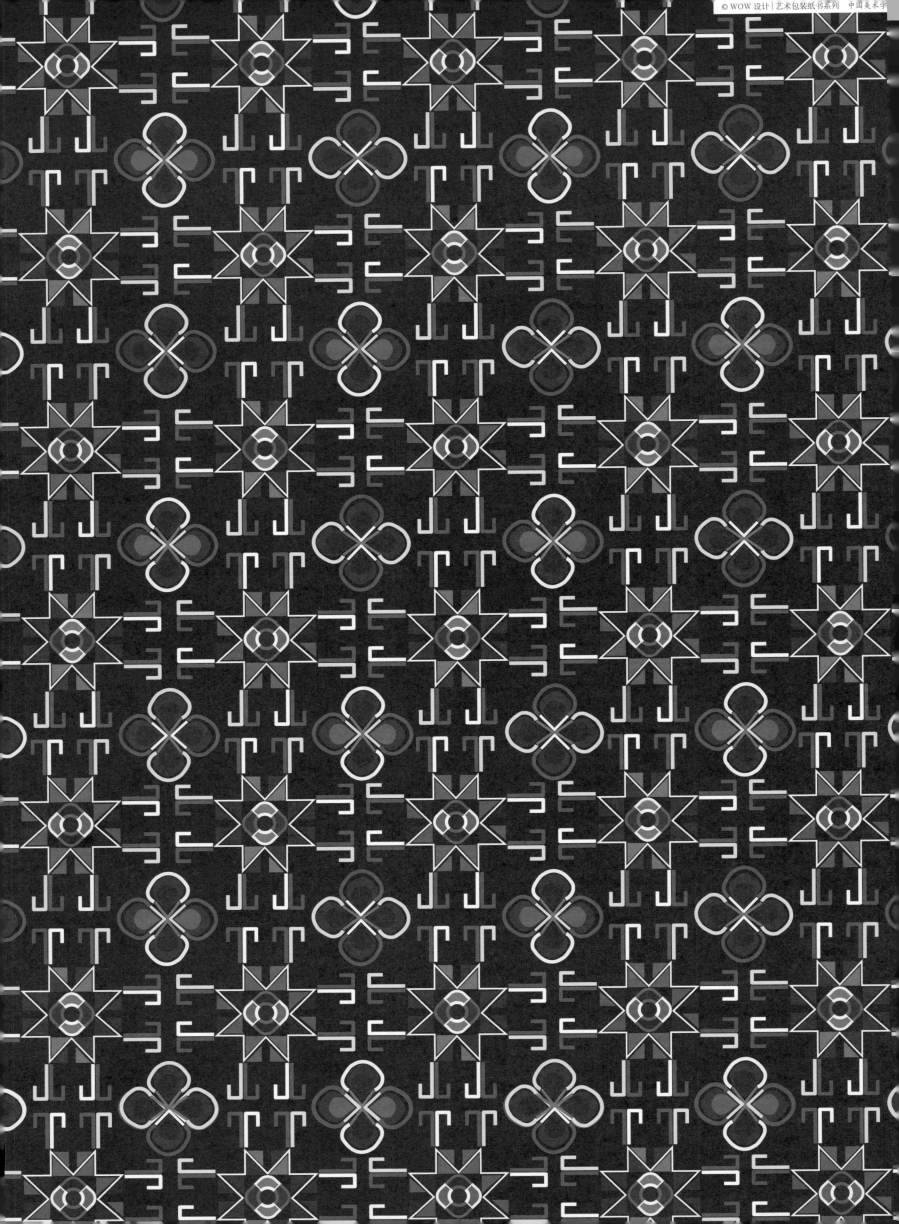

■

中国是多民族融合的大家庭，分布在不同地域的 56 个民族因文化、信仰的不同，织锦纹样呈现出不一样的风格。受到织造技艺的局限，这些纹样多以抽象几何的形式出现，以单独纹样连续不断地重复排列为主要形式，以日常生活中的动物、植物、天地属相、生活生产工具为原型提炼再设计后形成纹样，装饰在服饰、儿童背带、被面等上。

寓意吉祥、美好，是少数民族纹样的主要特征，来源于生活又高于生活，不同于实物的写实，令织锦纹样更抽象，但每一个纹样背后所承载的细腻和饱满，传达了各民族的审美情趣和民族信仰。蝴蝶长寿花传达了对美好生命延续的期盼，阳雀象征万物复苏、春天到来，太阳、月亮纹是对自然敬畏的表达，通常装饰在孩子的背带上，以祈佑其健康成长。但因不同民族文化背景的不同，相同纹样被赋予不同的含义，如八角星纹被认为是光芒四射的太阳、无际的天空。我们在惊叹这些纹样之美的同时，也不得不感慨当地妇女的设计巧思，她们将生活中的点滴美好和对美好生活的期盼以丝绸、布帛等为媒介，以纹样的形式记录下来，使其得以流传。

■

As a large united multi-national state, China is composed of 56 ethnic groups, whose cultures and beliefs differ largely, leading to different types of brocade patterns. Due to the limitation of weaving techniques, most patterns appear in form of abstract geometry, featuring continuous and repeated individual patterns. The animals, plants, signs of heaven, earth and 12 zodiac, productive tools, and life utensils are extracted and redesigned into patterns, and used as decorations on clothing, suspenders for kids, quilt cover, etc.

Patterns on silk fabrics of ethnic minorities in China often symbolize good fortune and beautiful life, which come from life but transcend life, and describe the real thing but deviate from the reality. Though seemingly abstract, each brocade pattern conveys the aesthetic interests and national beliefs of the ethnic group. The butterfly and Kalanchoe blossfeldiana pattern expresses the desire for a better life. The cuckoo bird symbolizes the resurrection of everything and the arrival of spring. The sun and moon pattern shows reverence for nature. These patterns are often seen on kids' suspenders to bless their health. However, due to the different cultural backgrounds of ethnic groups, the same pattern may have different meanings. For example, the octagonal star pattern could be considered as the shining sun and the endless sky. While marveling at the beauty of these patterns, we are also impressed by the ingenuity of local women in creating these designs on silks, cloths, etc. to document the beauty in their lives and their aspirations for a better life in the form of patterns and pass them down for generations.

WOW DESIGN

WOW 设计为中国美术学院出版社旗下品牌,特别联合中国美术学院设计专业团队以及国内外前沿插画家、艺术家倾力原创设计的具有高度艺术设计感的包装纸书系列,以及出版具有高度艺术品质的生活美学类图书及文创周边产品,以艺术、生活、图书为核心,以创造美、传递美为己任,以期一起发现和创造美好生活。

图书在版编目(CIP)数据

琐纹 / 赵丰主编;安薇竹编 . -- 杭州 : 中国美术学院出版社 , 2020.8
　(WOW 设计 . 艺术包装纸书系列)
　ISBN 978-7-5503-2265-3

　Ⅰ.①琐… Ⅱ.①赵… ②安… Ⅲ.①包装纸－设计 Ⅳ.① TS761.7

中国版本图书馆 CIP 数据核字 (2020) 第 089239 号

WOW 设计 | 艺术包装纸书系列　琐　纹　赵丰　主编　安薇竹　编

责任编辑	邓秀丽
封面设计	唐　泓
图片处理	孙培彦
责任校对	杨轩飞
责任印制	张荣胜

出 品 人	祝平凡
出版发行	中国美术学院出版社
地　　址	中国 · 杭州市南山路 218 号 / 邮政编码:310002
网　　址	http://www.caapress.com
经　　销	全国新华书店
制　　版	浙江新海得宝图文制作有限公司
印　　刷	杭州捷派印务有限公司
开　　本	787 毫米 ×1092 毫米　1/8
印　　张	12
字　　数	240 千
版　　次	2020 年 8 月第 1 版
印　　次	2020 年 8 月第 1 次印刷
书　　号	ISBN 978-7-5503-2265-3
定　　价	88.00 元

如有印刷装订质量问题,请联系出版社市场营销部调换。